A FIELD GUIDE TO

Harlequins

A FIELD GUIDE TO

Harlequins

and Other Common Ladybirds
of Britain and Ireland

HELEN BOYCE

PELAGIC PUBLISHING

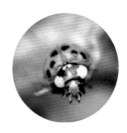

Published by Pelagic Publishing
PO Box 874
Exeter
EX3 9BR
UK

www.pelagicpublishing.com

*A Field Guide to Harlequins and Other Common
Ladybirds of Britain and Ireland*

ISBN 978-1-78427-244-9   *Paperback*
ISBN 978-1-78427-245-6   *ePub*
ISBN 978-1-78427-246-3   *PDF*

A CIP record for this book is available from the British Library

Cover photographs: Harlequin ladybirds

Top row, left to right: 7-spot, Pine and Adonis' ladybirds

Middle row, left to right: 7-spot, Striped and Cream-streaked ladybirds

Bottom row and right: Harlequin ladybirds

Contents

Foreword

Ladybirds are captivating creatures, and in this book Helen Boyce has conveyed this fact with eloquence and care. Her enthusiasm and expertise shine through on every page, as do her beautiful photographs. From a Bryony ladybird larva shedding its skin to the sequence of colour changes seen in a newly emerged 14-spot ladybird, the images here are fascinating and informative. Moreover, readers are sure to appreciate the clear and plain-English explanations in Helen's text, which avoids getting bogged down in unnecessary technical detail.

Learning to identify ladybirds is not always easy. There are a number of species that trick us with their variable colour patterns and defiance of the rules of where we expect to find them. In sharing her experience of mastering the identification of Harlequin ladybirds, Helen reveals the key features of other species too. Her approach allows the identification of these species in myriad ways. This will provide a fantastic start for those who are new to observing ladybirds, but will also prove insightful for more experienced recorders.

There are so many reasons to coordinate a biological recording scheme such as the UK Ladybird Survey. It is exciting to see records arrive throughout the year from across the UK; with them we get an almost instantaneous overview of the whereabouts of these amazing beetles. This book will undoubtedly help in such efforts. Knowing that ladybirds are being seen and appreciated by so many people is highly rewarding and it is always a pleasure to hear from recorders and to share stories of ladybird encounters. It is both the ladybirds and the people that make the UK Ladybird Survey what it is – an absolute joy.

We congratulate Helen on an excellent contribution to the study of ladybirds. The attention to detail is impressive. We wish all the readers of this book many and varied ladybird days.

Helen Roy and Peter Brown
UK Ladybird Survey

Introduction

When I was first attempting to learn which ladybirds were Harlequin ladybirds, I looked around for a handy book that would give me lots of helpful information and instructive photographs – but no such book existed. So, I have decided to put together the guide that I wish had been available when I was learning.

I am most often asked: 'How do you know it's a Harlequin?' Now that I work to help others learn about and recognise both the Harlequin ladybird and the other ladybirds found in Britain and Ireland, I am going to try to answer that question in this guide.

There are 47 species of ladybirds in Britain and Ireland, but only 26 of them – including the Harlequin – are conspicuous, looking rather like the classic ladybird of our childhoods.

The Harlequin ladybird (*Harmonia axyridis*) gets its name from its bright colours and variety of patterns, but due to its wide distribution it is known by many other names in different parts of the world, including the Asian or Halloween ladybug or lady beetle, the Japanese, Southern, Pumpkin or Multicoloured ladybird, to name but a few.

This book will help you to recognise a Harlequin ladybird and to tell it apart from the other 25 colourful ladybirds currently found in Britain and Ireland.

Above: Harlequin larva (*left*) and pupa (*right*)

Above: Harlequin ladybirds

The Harlequin ladybird: how and when it arrived

The Harlequin ladybird (*Harmonia axyridis*) was introduced from Asia into North America in the 1980s to control aphids that were feeding on farmers' crops. However, the Harlequin quickly spread across the USA to become the most common ladybird there. So, the Harlequin is an example of a pest-control species that has itself become a pest.

Between 1982 and 2003 the Harlequin ladybird (*above*) was introduced to several European countries, again as a form of biological pest control.

The Harlequin ladybird was first recorded in Britain in 2003 and had firmly established itself by 2005. It spread rapidly, at a rate of more than 100 kilometres per year, and is now very common and widespread over most of England and Wales, with increasing numbers also recorded in Scotland and Ireland. It was unintentionally introduced to Britain, probably in a number of ways. Some Harlequins (*below*) might have flown across the channel, while others were possibly transported with goods and produce. There are also accounts of Harlequin ladybirds being brought into the country accidentally, in people's luggage.

In 2005 the UK Ladybird Survey, a citizen science initiative involving online recording, was launched to encourage people across Britain to help track the spread of the Harlequin. Tens of thousands of people provided records of ladybirds, and this has created an invaluable dataset for research and understanding.

Harlequins, it has been said, are the most invasive ladybird on earth. They may live longer than most other ladybird species, and they have a greater breeding capacity and a longer breeding season.

Not only do they out-compete other ladybirds by eating the food that they need, they also have a tendency to actually eat other ladybirds. Although the long-term impact on the native species is not yet known, since the Harlequin ladybird arrived on our shores, there has been a decrease in the numbers of several other species of ladybirds, and it is believed that the Harlequin is the main reason why the 2-spot ladybird is now so scarce.

Above: A Harlequin gorging on a blackberry

The descriptions and photographs in this guide will help you to recognise the Harlequin ladybirds found in Britain and Ireland, in all their variations.

Where and when to find Harlequin ladybirds

The Harlequin ladybird (*below*) can be found in a wide variety of
habitats, though it tends to be more prevalent in urban and suburban
localities. It can often be found in parks and gardens, and in or on
buildings, as well as on arable crops and in woodlands. This species
thrives on deciduous trees and shrubs such as limes, maples, birches
and roses as well as a wide variety of herbaceous plants, but seems to
favour nettles, thistles, brambles, cow parsley and hogweed as well as
ornamental garden plants.

The best time to find ladybirds, including Harlequins (*above and below*), is on warm, dry and sunny days. Harlequins can be encountered in a wide range of places, for example on the exterior of buildings, on leaves, stems, branches or tree trunks. On days when it is raining or dull, you may have to look a bit harder, but peer under leaves and flowerheads and you will often find them sheltering. September and October are usually good months to find the adults.

In Britain, the Harlequin ladybird (*above*) is widespread. It is most frequently found in the south and southeast of England but is also fairly common in the Midlands and parts of Wales. It occurs to a lesser extent in northern England, Scotland, Ireland, the Isle of Man and the Isles of Scilly.

Harlequins often spend the cold winter months clustered together in an 'aggregation' (*left*). They tend to choose elevated positions, and in October and early November aggregations of Harlequins can often be found on or in buildings, for example in upstairs window frames (even double-glazed ones), lofts and church porches.

Although they are not a danger to us, in spring they can be a nuisance as they try to find their way out of the house again. They give off an unpleasant smell, can sometimes stain curtains and carpets, and can even very occasionally cause an allergic reaction.

Above: Harlequin ladybirds

Characteristics of a ladybird

The colourful and conspicuous ladybird is a small to medium-sized beetle. Some are only 3 millimetres long, while others can be up to 8.5 millimetres in length. They have large compound eyes and their antennae are slightly clubbed. The head can be partly withdrawn under the hard, protective plate known as the pronotum which covers the soft thorax. They have short legs which they can retract

Pronotum (hard plate protecting the thorax)

Elytra (wing cases)

Head

into depressions under their body. Adult males and females look similar, although males tend to be slightly smaller.

The abdomen is covered by the outer wing cases, known as the elytra, which are actually the hardened, often brightly coloured forewings. Hidden under the outer wing cases are the delicate membranous hindwings, carefully folded up into a 'Z' shape when not in flight. Owing to the intricate folding of these long hindwings, it takes time for the ladybird to unfold them and take off. Just before take-off, the elytra swing outward, allowing the ladybird's hindwings to unfold. Once they start beating, the ladybird launches itself into the air. The elytra are important for providing both lift and steering in flight.

Above: An Eyed ladybird preparing for flight

Above: A 7-spot ladybird preparing for flight

Contrary to popular belief, the number of spots on a ladybird has no connection with how old it is – and in some cases there is little connection even with the name of the ladybird, as the number of spots can vary greatly within a species.

Left: The 24-spot ladybird usually has 20 spots, but can have anything between 0 and 26 spots

The 10-spot ladybird (*above and left*) is a very variable species that can be found in a variety of different combinations of colours and patterns – for example black, purple, dark brown, yellow, orange or red background colours with spots, a grid-like 'chequered' pattern, or yellow or red shoulder flashes.

Some species of ladybird show only slight variation in colour and markings – for example, the Orange ladybird (*right*), the 22-spot ladybird (*below left*) and the Bryony ladybird (*below right*).

There are two species of ladybird that show almost no variation in colour or markings at all – the Cream-spot ladybird (*below left*) and the Kidney-spot ladybird (*below right*).

The lifecycle of a ladybird

Eggs

Female ladybirds lay batches of between 1 and 100 eggs at a time (*right*). They lay their eggs near a food supply (*below*) in late spring and early summer. However, the Harlequin can breed throughout the summer, and as a result there can be two generations a year rather than just one.

The small, elongated ovoid eggs are usually laid upright and in tight clusters. They are smooth in texture, and can be anything from off-white to dark orange. If conditions are right, a female may lay over 1,500 eggs in her 14- to 15-month lifetime.

Due to their longer breeding season and the fact that they can live for longer than other species, an individual female Harlequin may produce many more eggs in her lifetime than any other ladybird. The eggs are about 1.2 millimetres long and take about five days to hatch, depending on temperature.

 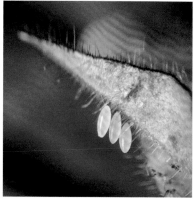

Above: Bryony ladybird eggs (*left*); 14-spot ladybird eggs (*right*)

Above: Cream-streaked ladybird eggs (*left*); Cream-spot ladybird eggs (*right*)

Larvae

After 2–10 days, depending on temperature (the warmer it is, the sooner they are ready to hatch), the tiny larvae hatch out of

the eggs (*left*). They usually eat the empty eggshells, and often any unhatched eggs are eaten as well. Eating the unhatched eggs gives the tiny larvae a far greater chance of survival, as they will die unless they find food within a day or two. However, as the eggs are usually laid in close proximity to their food, young larvae of predatory ladybirds (for example 2-spot, 7-spot, Kidney-spot, Adonis' and Harlequin ladybirds) usually find aphids or other similar food close by. Equally, ladybirds that are mildew eaters (for example 22-spot and Orange ladybirds) and plant-eating ladybirds (for example 24-spot and Bryony ladybirds) are usually standing on, or very near to, their required food.

Left: A one-day-old Cream-spot larva, barely bigger than a pin head at just 1.3 mm long, successfully catches its first aphid

The growing larva sheds its skin (moults) three times over a three- to six-week period and then sheds its skin for the fourth and final time when it pupates, so it goes through four instar stages.

The larvae tend to eat the same food as the adults of the same species. A late-instar aphid-feeding larva and an adult ladybird might both eat in excess of 60 aphids a day. The larvae feed voraciously, and their elongated bodies rapidly increase in size.

Above: Third-instar 14-spot larva (*top*) and Harlequin larva (*bottom*) moulting (for the third time)

Above: A third-instar Bryony ladybird larva sheds its skin (for the third time): the start of the moult (*left*), two minutes later (*centre*), and 18 minutes later (*right*)

Above: Striped ladybird larva (*left*), and shed skin
from a Striped ladybird larva (*right*)

Above: Cream-streaked ladybird larva (*left*), and shed
skin from a Cream-streaked ladybird larva (*right*)

The larvae of different species of ladybird vary in colour and markings, and during the first two instar stages it can be difficult to tell them apart. By the third or fourth instar they become easier to identify as their distinguishing markings and colours are more fully developed.

Ladybird larvae of Britain and Ireland

Harlequin ladybird
(*Harmonia axyridis*)

Cream-streaked
ladybird (*Harmonia
quadripunctata*)

Eyed ladybird
(*Anatis ocellata*)

7-spot ladybird
(*Coccinella
septempunctata*)

Scarce 7-spot ladybird
(*Coccinella magnifica*)

11-spot ladybird
(*Coccinella
undecimpunctata*)

10-spot ladybird (*Adalia
decempunctata*)

2-spot ladybird
(*Adalia bipunctata*)

Adonis' ladybird
(*Hippodamia variegata*)

Cream-spot
ladybird (*Calvia
quattuordecimguttata*)

14-spot ladybird
(*Propylea
quattuordecimpunctata*)

13-spot ladybird
(*Hippodamia
tredecimpunctata*)

16-spot ladybird
(*Tytthaspis
sedecimpunctata*)

18-spot ladybird
(*Myrrha
octodecimguttata*)

Larch ladybird
(*Aphidecta obliterata*)

Hieroglyphic
ladybird (*Coccinella
hieroglyphica*)

5-spot ladybird
(*Coccinella
quinquepunctata*)

Striped ladybird
(*Myzia
oblongoguttata*)

Water ladybird
(*Anisosticta
novemdecimpunctata*)

Bryony ladybird
(*Henosepilachna
argus*)

24-spot ladybird
(*Subcoccinella
vigintiquattuorpunctata*)

Kidney-spot
ladybird (*Chilocorus
renipustulatus*)

Pine ladybird
(*Exochomus
quadripustulatus*)

Heather ladybird
(*Chilocorus bipustulatus*)

22-spot ladybird
(*Psyllobora
vigintiduopunctata*)

Orange ladybird
(*Halyzia
sedecimguttata*)

HOW TO RECOGNISE A HARLEQUIN LARVA

Harlequin larvae have distinctive fleshy, branched spines known as 'scoli' (*left*). Until the newly hatched larvae have shed their skins twice and are third-instar larvae, they are very hard to identify. In the third and fourth instars, the orange markings are more pronounced and identification becomes easier.

Above: Newly hatched 2 mm-long Harlequin larva (*left*),
4 mm-long second-instar larva (*centre*), 7 mm-long third-instar larva (*right*)

Left: The colouration and markings of the 10 mm-long fourth-instar Harlequin larva are bolder and more extensive than on earlier instars

Above: Harlequin larvae have fleshy, branched spines known as scoli

Right: The 7-spot ladybird larva has fine hairs rather than scoli and two pairs of orange-coloured segments down both sides as well as orange markings behind the head

The only larvae that are likely to be confused with the Harlequin larvae are those of the Cream-streaked ladybird.

Above: Cream-streaked ladybird larva (*left*) and Harlequin ladybird larva (*right*), showing the differences in their markings

Both the Harlequin and the Cream-streaked ladybird larvae have a row of branched orange scoli (as opposed to bristles or unbranched spines) running down each side of the body. However, the Cream-streaked ladybird larva (*above left*) only has four orange scoli on each side of its body, whereas the Harlequin larva (*above right*) has five orange scoli on each side, with the two lines nearly meeting at the top of the body (two 'L'-shaped lines) and four central orange-yellow scoli between the two lines nearer the tail end. In both species, the fully grown larva is about 1 centimetre long.

Pupae

When the fourth-instar larva is fully grown, it attaches itself to a surface (for example a leaf or stem) with the help of anal pads. It then hunches over (*right*) and after a day or so spent in this pre-pupal stage, its larval

 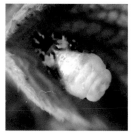 

skin splits for the fourth and final time, revealing the pupa, either fully – as in the case of the Harlequin ladybird (*above*) – or partially, as seen, for example in the 24-spot ladybird (*near right*) and the Bryony ladybird (*far right*).

Above: Harlequin pupa (*left*), 7-spot pupa (*centre*),
and the much smaller 22-spot pupa (*right*)

With some species of ladybird that shed their larval skin completely, the remains of their old skin can still be seen throughout pupation. It's as if they have dropped their overalls and left them around their ankles! The remains of the skin at the base of the pupa can be a very helpful identification tool.

Instead of completely splitting their old larval skins, some species (such as the Kidney-spot, *below left*), pupate within their partially split larval skin, which acts as a 'cradle' until the adult finally emerges (*below right*).

The pupae are vulnerable to predation, but they can defend themselves to some degree by 'flicking up' their front, unattached end (Harlequin pupa, *right*) in an attempt to startle a potential predator or parasite.

The processes of metamorphosis that take place while the ladybird is a pupa result in a remarkable transformation. The larval body and many of the organs are broken down to reform the new organs of the adult ladybird. Depending on the temperature, this process can take a few days to a couple of weeks, at which point the adult ladybird is ready to emerge.

Above: 10-spot ladybird pupa (*left*), 14-spot ladybird pupa (*right*)

HOW TO RECOGNISE A HARLEQUIN PUPA

The orange pupa of the Harlequin ladybird (*left*) is relatively easy to recognise from the discarded skin that is left at the base of the pupa, as this old skin still has the distinctive forked and branched black and white scoli that were on the larva.

Above: Harlequin pupa (*left*) with black and white scoli visible at the attached base; a 7-spot pupa (*right*) does not have the forked scoli at the base, just bristles and old skin

Apart from the Harlequin, the only other British ladybird pupa that has the distinctive feature of the scoli at its base is the Cream-streaked ladybird pupa (*opposite, below*), but the colouration and markings of the two pupae are different.

Above: Harlequin ladybird pupae and an empty Harlequin pupal case (*right*)

The Harlequin pupa is orange with pairs of squarish black markings down each side and a black patch part-way down, and the black and white scoli of the shed larva skin remain visible at the base of the pupa.

The Cream-streaked ladybird pupa is light beige with six rows of black spots and dashes. As with the Harlequin pupa, the black and white scoli on the shed larva skin remain visible at the base.

Above: Cream-streaked ladybird pupae

Adult ladybirds

THE NEWLY EMERGED ADULT

After a few days the new adult emerges from its pupal case. Initially, it sits on the now empty case for a while, allowing both its membranous flight wings to expand and its wing cases (elytra) to expand and dry.

At first the freshly emerged ladybirds have a slightly 'orange peel' matt appearance to them, but they soon look very glossy. The soft wing cases appear translucent, pale yellow, orange or white. This is the case even with species that will eventually be predominantly black. For this reason, newly emerged adults can be very difficult to identify.

Above: A 7-spot ladybird about 1 hour after emerging, and how it looked 24 hours later

Above: This Harlequin took 8 minutes to emerge from its pupal case before resting and expanding its wings for 45 minutes (*middle right*). The photos in the bottom row show the colour developing 6 hours and 18 hours later.

Unlike the markings on the wing cases, the markings on the pronotum are often fully developed by the time the new adult emerges. (The pronotum is the hard, protective plate that covers the soft thorax.)

Left: A newly emerged 22-spot ladybird with a fully developed pronotum

Above: Sequence showing the darkening colours of a 14-spot ladybird that has just emerged from its pupal case: after 15 minutes, 2 hours, and 3 days

Above: Harlequin ladybird: newly emerged, after 3 hours, and after 12 hours

Above: Newly emerged 24-spot ladybird: emerging, 6 hours, 24 hours

After an hour or so the ladybird is ready to fly, but the full colour of the wing cases (and underside) can take anything from a few hours to several days (and sometimes much longer) to develop. In the red species, the shade of red may continue to darken throughout the life of the ladybird.

There is considerable variation in the timing of emergence from pupae in different species. This variation is a consequence of a variety of factors, weather (temperature) and food availability being the most important. However, the majority of ladybird species emerge between mid-July and mid-August, then feed up before dispersing to their overwintering sites.

Right: Two 7-spot ladybirds basking in the late-summer sun with a 16-spot ladybird

OVERWINTERING

In September and October, ladybirds select their overwintering sites. Some simply find a slightly sheltered spot such as a curled-up leaf or hollow plant stem to pass the inhospitable winter months. Other ladybirds use exposed tree trunks or cracks in bark, or shelter near the ground, for example in clumps of grass, under leaf litter or between stones. During warm spells they can temporarily come out of dormancy to eat and drink.

Above: 7-spot ladybird (*left*) and Harlequin ladybird (*right*) using dead leaves for shelter

Most species of ladybird have a tendency to group (aggregate) together in winter and it is not uncommon to find mixed-species groups. Aggregating may offer additional protection from harsh weather, and it might also help them find a mate more quickly when they re-emerge in the spring.

Opposite: Harlequin ladybirds (*left and middle row right*) and 7-spot ladybirds (*top and bottom right*) using dead leaves, stems and thistle heads as shelter

Above: A small aggregation of 16-spot ladybirds huddled under a branch on a willow sapling (*left*); a 16-spot ladybird sheltering under a splinter of wood (*right*)

Some species of ladybird choose to overwinter on trees, preferring smooth bark and thin trunks and branches. They position themselves directly below an overhanging twig, branch or knot, usually tending to select the side of the tree that offers the most shelter from the prevailing winds.

Left: A well-camouflaged Cream-streaked ladybird sheltering between needles in a Scots pine tree

Above: An aggregation of 16-spot ladybirds on a 25 × 25 mm tree stake

The ladybird expert Michael Majerus put forward the fascinating theory that in October, some species of ladybird seem to be able to predict the severity of the winter ahead and will apparently select their overwintering site accordingly. He suggested that if ladybirds predict that it will be a mild winter, they choose a relatively exposed site, whereas if the winter ahead is going to be a severe one, they seek a more sheltered position.

Right: A 7-spot ladybird sheltering amongst frosted thistle leaves

SPRING EMERGENCE

Come spring, and in response to lengthening days and warmer temperatures, the ladybirds start to emerge from their wintering places to go in search of food and a mate. On warm sunny spring days, when they are at their most active, mating ladybirds can be a common sight.

Mating ladybirds (*clockwise from top left*):
22-spot, Kidney-spot, Bryony, Harlequin

Protection, predation and parasites

The ladybird's strong pattern of two or more bright contrasting colours is thought to warn predators of its toxicity and unpleasant taste. When crushed, ladybirds emit a strong and unpleasant smell, and when threatened they generally 'play dead' by withdrawing their legs and antennae close to their bodies and keeping still. Most species of ladybird also exude a noxious yellow fluid, known as 'reflex blood', which oozes from their leg joints and forms droplets at the edge of the pronotum and wing cases to deter predators.

Right: Harlequins reflex bleeding as a defence mechanism

Owing to their toxic properties, reflex bleeding and other defence strategies, ladybirds are seldom attacked by predators. For many species of ladybird, including the Harlequin (*left*), their strong, bright colouring and their unpleasant smell give them additional protection. However, some birds (for example house martins and tree sparrows) and invertebrates (for example spiders and other species of ladybird) do attack and eat ladybirds. The Harlequin has higher levels of toxicity and seems to be even more protected against predation than many other species of ladybird, and incidents of Harlequins being eaten by birds, for example, have only rarely been reported.

Left: Harlequin ladybird caught in a spider's web

There are a number of parasites that prey on ladybirds and use them as hosts, including parasitic flies and wasps.

The most common parasitic fly that preys on ladybirds usually lays its eggs between the legs of pre-pupal larvae or on the underside of newly formed pupae. These eggs then hatch within a few hours and burrow into the host, where they stay until they are ready to emerge and pupate.

The most common of the parasites is the wasp *Dinocampus coccinellae*. The female wasp lays an egg in the ladybird's body, and when the egg hatches the wasp larva feeds on fat reserves and nutrients from the body of the host. Just before the larva emerges from the ladybird, it partially paralyses but does not kill its host, emerges and then spins a cocoon under the ladybird's legs. The paralysed ladybird is attached to the cocoon and the cocoon is protected by the ladybird's body and bright warning colours until the adult wasp emerges about a week later. The ladybird usually then dies of thirst, starvation or fungal infection.

Right: Partially paralysed 7-spot ladybird with a cocoon of the parasitoid wasp *Dinocampus coccinellae* between its legs

Early studies showed that the Harlequin was initially less susceptible to attack than other species of ladybird. However, it appears that parasites have begun to adapt and are now targeting Harlequins too.

There is a fungal infection called *Hesperomyces virescens* which is considered to be a sexually transmitted disease, transferred from ladybird to ladybird when they are in close contact, for example during mating. With Harlequins, infection is also socially transmitted, for example as they overwinter, huddled close together in aggregations. Although ladybirds don't die from this infection, it may shorten their lifespan and reduce their ability to lay as many eggs as normal. You can easily see the yellow fruiting bodies of *H. virescens* with the naked eye on ladybirds – it looks like a spikey yellow crust, usually starting on the back end of the ladybird.

Above: Two Harlequin ladybirds with the fungal infection *Hesperomyces virescens*

Food

Of the 26 conspicuous ladybird species found in Britain, 21 are predatory, and for the vast majority of these, aphids are the main food source. The remaining five species feed on either mildew or plants.

Above: Harlequins eating blackberries (*top left*), larvae (*right*) and a fly (*bottom left*)

Above: Harlequin ladybirds

When food sources are low, some ladybirds will eat the eggs and larvae of other ladybird species. This is known as 'intraguild predation'. Harlequins tend to do this more often, and it is one important way in which they are a threat to the other species of British ladybird.

The Harlequin ladybird (*above*) has a voracious appetite and is a very capable and proficient hunter. Unlike most other British ladybirds, which have quite restricted diets, Harlequins will readily eat a wide variety of foods including aphids, soft fruit and pollen as well as other insects, including other ladybirds and butterfly and moth eggs and caterpillars.

Right: A Harlequin larva eating another Harlequin larva that was about to pupate and therefore unable to run away

How to recognise a Harlequin ladybird

1. Introduction to Harlequin colour forms

Adult Harlequins are one of Britain's largest ladybirds, at between 5 and 8 millimetres in length. They are about the same size as the 7-spot ladybird. Some Harlequins are black with red spots and some are red, orange or yellow with black spots. Their appearance can vary widely, which can make them hard to tell apart from other ladybirds.

In their native Asia, more than 200 different colour patterns have been recorded, although most of them can be categorised into just four main colour forms. There are some other Harlequin forms, and these are described on pages 93–95.

In Britain, the range of Harlequin colour forms (f.) is limited to just three:

f. *succinea* – red, orange or yellow background colour with between 0 and 19 black spots (*below*). This is by far the most common form in Britain.

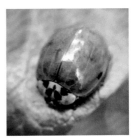  

f. *conspicua* – black background colour with two large red, orange or yellow spots/patches, one on each side (*below*).

f. *spectabilis* – black background colour with four red or orange spots/patches, two on each side (*below*).

However, there is also a wide range of different markings and patterns within these three forms. The following sections of this book will focus on clarifying the various identification features of the Harlequin ladybird, and on the differences between Harlequins and the other ladybird species of Britain and Ireland.

2. Only Harlequins have two 'shoulder spots'

Of the larger (5–8 mm) red, orange or yellow (red/orange) ladybirds, Harlequins are the only ones that have two black spots ('shoulder spots') right at the front of the wing cases, just behind the pronotum. These two 'shoulder spots' are always relatively close together and lie parallel to the edge of the wing case/pronotum. A few of our other red/orange natives also have two black spots quite near the front of their outer wing cases, but the top spot is always higher up and further back. The Harlequin's 'shoulder spots' are close together and are an easy way to recognise it, even from a distance.

Above: Harlequin ladybirds with two 'shoulder spots'

Above: Harlequin ladybirds (5–8 mm) with two 'shoulder spots'

Right: The 10-spot ladybird (3.5–4.5 mm) is smaller and only has one 'shoulder spot'

3. Harlequins *never* have white 'angel wings'

Some species of British ladybird, for example the 7-spot ladybird (*below*), have a black spot on the centre line of the elytra, just behind the pronotum, which has distinctive white patches each side. Although it is a bit of a fanciful idea, try and think of this as looking like an angel with white 'angel wings', with a round black body and a little black head in between the white wings.

Above: Two examples of British ladybirds with white 'angel wing' markings: Adonis' ladybird (*left*) and Hieroglyphic ladybird (*right*)

Seven different species of British ladybird have distinct white 'angel wing' markings, but Harlequins never do.

Right: The Striped ladybird has elongated white 'angel wings'

If you see a ladybird with white 'angel wing' markings (*right*), then you know it is *not* a Harlequin.

4. Harlequins are 5–8 millimetres long

Female ladybirds are usually, but not always, larger than the males. Harlequin ladybirds are between 5 and 8 millimetres in length. However, that 3-millimetre variation can make quite a big difference to the way they appear. They are larger than many British ladybirds, and to help with visualisation, that is roughly the size of a split pea or a garden pea cut in half.

Middle left: The Harlequin ladybird (*top*) and the 7-spot ladybird (*bottom*) can appear very similar in both size and shape

Bottom left: The Harlequin ladybird (5–8 mm) on the left is considerably larger than the tiny Pine ladybird (3–4.5 mm) on the right

Although initially the Harlequin (*bottom left*) looks very similar to the spotted form of the 10-spot ladybird (*bottom right*), two obvious things tell them apart – the 'shoulder spots' (one or two) and their size. The 10-spot ladybird is only 3.5–4.5 millimetres in length and has just one 'shoulder spot', whereas the Harlequin ladybird is 5–8 millimetres (about the size of half a garden pea) and has two 'shoulder spots'.

When trying to identify any ladybird, size is really important.

5. The only black ladybird that has full, round white 'cheeks' is the Harlequin

The easiest way to tell the difference between a black Harlequin (*above*) and other black ladybirds is that black Harlequins usually have distinctive, full round white 'cheeks', whereas other black ladybirds lack this feature. Of course, these areas of white aren't really cheeks at all, as they are actually on the pronotum (the hard, protective plate that

Above: Black Harlequin ladybirds with typical white 'cheeks'

covers the soft thorax and the head when it is tucked in). With the head tucked in, these round white areas do look rather like cheeks!

Apart from the Harlequin, there are three black British ladybirds: the Pine ladybird (3–4.25 mm) (*below left*), the Kidney-spot ladybird (4–5 mm) (*below right*) and the Heather ladybird (3–4 mm). These diminutive ladybirds usually have no white on them at all, although occasionally they do have a really tiny crescent of white 'cheek' on the pronotum.

Some ladybirds that are usually red/orange also appear in melanic (black) forms, for example the 2-spot (4–5 mm) and 10-spot ladybird (3.5–4.5 mm). Melanic forms of 2-spot and 10-spot do have some white on their pronotum (a 'crescent' of white), but it is only black Harlequins that have the distinctive full, round white 'cheeks'.

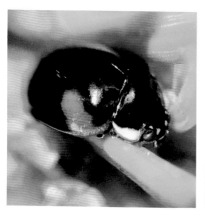

Above: In the melanic form of the 10-spot ladybird the red patches extend to the very edges of the elytra

The Hieroglyphic ladybird (4–5 mm), which is usually reddish-brown, is the only other British ladybird that commonly appears in melanic form (*left*), and it has distinctive triangular-shaped 'cheeks' (as well as small white 'angel wing' patches, just behind the pronotum).

Above: Although the majority of black Harlequins have full, rounded white 'cheeks', they don't always, as this selection shows

Sometimes a red/orange Harlequin may appear, at first glance, to be black. This happens most often to ladybirds emerging in autumn once the temperature has dropped. The background is darker and the spots merge more than usual. But if you look closely you can usually still see the telltale black spots, specifically the two 'shoulder spots'.

Above: Two very dark red/orange Harlequin ladybirds

Above: Although these two ladybirds look very similar, the Pine ladybird on the left is only 3–4.5 mm long with no white markings, whereas the Harlequin on the right is much larger at 5–8 mm long and has white 'cheeks'

If you come across one of the black British ladybirds other than a Harlequin that has small white crescent-shaped 'cheeks', it should be possible to identify it not only by the size and shape of the white 'cheeks', but also by its other markings, its leg colour and its size.

Left: Pine ladybird with a very thin white crescent 'cheek' on its pronotum and the normal, distinctive red front 'speech mark' spot

Note: This white 'cheek' rule of thumb only applies to black ladybirds. It does not apply to red/orange ladybirds.

6. The Harlequin's pronotum is always black and white/cream

The Harlequin ladybird's pronotum is always black and white or black and cream (*below*).

In contrast, some British ladybirds have a pronotum that is the same colour as the outer wing cases – for example the 16-spot ladybird (*below left*), the 24-spot ladybird (*below centre*) and the Water ladybird (*below right*).

7. The Harlequin's pronotum markings

Red/orange Harlequins (5–8 mm) often have a distinctive black 'M' mark on their pronotum (*below*). The smaller 2-spot ladybird (4–5 mm) and larger, burgundy-coloured Eyed ladybird (7–8.5 mm)

are the only other British ladybirds likely to have 'M' pronotum markings; the 'M' markings on both of these species are chunkier than on the Harlequin.

Left: Eyed ladybird with chunky 'M' pronotum markings

Red/orange forms of Harlequin ladybirds can have a variety of patterns and markings on the pronotum (*above and right*), and sometimes the spots and markings can be very similar to the markings on the pronota of other British ladybirds.

Above: The pronotum of the Harlequin on the left and that of the 10-spot on the right look very similar. In order to identify which is the Harlequin, look first at whether or not it has two 'shoulder spots'. If it does, then it is a Harlequin. If it does not have two 'shoulder spots', then consider the size: Harlequins are 5–8 mm long, whereas 10-spot ladybirds are much smaller, only 3.5–4.5 mm

Another example of a ladybird that might be confused with the Harlequin is the Cream-streaked ladybird (*left*).

The Cream-streaked ladybird is 5–6 mm long and usually has 5–9 black spots on its pronotum, whereas the Harlequin is 5–8 mm long and usually has a maximum of five spots on its pronotum.

Above: Cream-streaked ladybird

The 7-spot is the ladybird that is perhaps most commonly confused with the Harlequin as it is about the same size (5–8 mm in length). However, the pronotum on the 7-spot (*below*) is always black with roughly square (rather than round) white 'cheeks'. (Note also the 'angel wings' on the 7-spot ladybird.)

Above: Harlequin ladybird

8. Harlequin spots and patterns

As already shown, in Britain and Ireland the range of Harlequin colour forms is limited to just three:

f. *succinea* (*below*) – red, orange or yellow background colour (red/orange) with between 0 and 19 black spots. This is by far the most common form found in Britain.

f. *conspicua* (*below*) – black background colour with two large red, orange or yellow spots/patches, one on each side.

f. *spectabilis* (*below*) – black background colour with four red or orange spots/patches, two on each side.

There is considerable variation in both pattern and colour within these three forms, especially in the red/orange *succinea* forms. With *succinea* forms of the Harlequin, there are up to 19 spots that are always in roughly the same positions – one central spot at the very front of the wing cases, then down each side there are two spots, then three spots, then three spots, then one spot at the end.

Variation in the markings is due to how many of the spots are visible, the size and strength of the spots, and whether or not they have merged together. The black spots can vary from big and bold through to very small and/ or very faint or even none at all. In some individuals the spots are clearly defined, whereas in others the spots are merged or fused to form different shapes and patterns (*right*).

Above: The black spots have merged considerably on both these Harlequins, but the two 'shoulder spots' are still visible. The Harlequin on the right has a darker background colour and is typical of adult Harlequins that emerge from their pupa in early autumn, once the weather has grown cooler

On some Harlequins the spots are distinct, bold and clear (*above left*), whereas on others the spots have merged to some degree or other, often making distinct patterns (*above right and below left*)

Note that the Harlequin on the left has only just landed on the leaf and has not yet had time to fold its membranous hindwings away

Occasionally the spots of the Harlequin merge together in such a way that it looks as if the ladybird has black bands or lines across it (*below*). However, on closer inspection you can see that these are simply spots that have fused or merged together.

With some Harlequins there appear to be no spots showing at all, though you might be able to notice one or two, perhaps near the outside of the wing case (*right*).

Occasionally, one or both of the two front 'shoulder spots' can't be seen at all, and sometimes they can only be seen very faintly (*above*).

Some Harlequins only have six spots – three each side (with or without other spots faintly visible) (*above*). One side of a ladybird can even look different to the other (asymmetrical), as in this Harlequin (*left*).

Above: Some forms of 10-spot ladybird (3.5–4.5 mm) (*left*) can appear to have almost identical markings to some Harlequins (5–8 mm) (*right*). However, the 10-spot ladybird is smaller, has a maximum of only one 'shoulder spot' on each side, and most likely has more spots on its pronotum

Black Harlequins can have either two or four red/orange spots, and they always have some degree of white 'cheeks' on the pronotum (*above and on next page*).

Above: Harlequin ladybirds

9. Harlequins do *not* have white spots on their wing cases

If you come across a ladybird with white or cream spots or dashes on its wing cases, then you know it is not a Harlequin, because Harlequins do not have white markings of any description on their wing cases.

Several British ladybirds have white 'angel wing' markings – for example the Hieroglyphic ladybird (*above left*) and the Adonis' ladybird (*above centre*). The Striped ladybird (*above right*) has white stripes and dashes.

Some British ladybirds have white spots – for example the Orange ladybird (*above left*) and the Cream-spot ladybird (*above centre*). 10-spot ladybirds can also have cream spots or patches (*above right*).

10. Harlequin leg colour

Although Harlequins are often described as having brown legs (*left*), the reality is that their legs can appear to be anything from light golden brown to almost black (*below*), but never jet black.

However, many other species of ladybird have unmistakably jet black legs, like the 7-spot ladybird shown on the right.

11. Harlequin underside colour

Both black Harlequins (*top row*) and red/orange Harlequins (*middle row*) have a two-tone underside. The outside edges tend to be orange-brown and the middle area black-brown. If you are able to see underneath, this can be a very helpful identification feature.

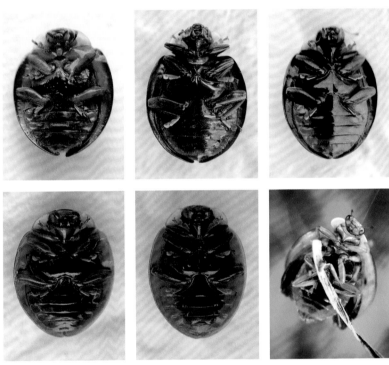

The underside of the similarly sized 7-spot ladybird, by comparison, is distinctly jet black all over (*right*).

12. Harlequins can have tiny rear 'pinched' dimples and ridges

Harlequins often have a pair of tiny 'pinched' dimples on the rear of their wing cases (known as the elytral ridge or transverse fold), and a tiny horizontal 'pinched' ridge (known as a keel). Sometimes these dimples are quite pronounced and at other times barely visible or not there at all. Although these dimples can also occasionally be found on a few other species, they can still be a helpful identification tool – although perhaps more so when viewing photographs rather than while out in the field.

Above: Two Harlequins with 'pinched' dimples and ridges

Above and right: Harlequins with
'pinched' dimples and ridges
(also known as elytral ridge/
keel or transverse fold)

13. Harlequins have a lip on the elytra

In common with some other species of ladybird (notably the much smaller Kidney-spot, Heather and Pine ladybirds, where the feature is very pronounced), the Harlequin often has a slight lip, or rim, on the outer edges of the elytra.

Above: Harlequins with rear 'pinched' dimples and the characteristic slight lip or rim on the outer edges of the elytra

Above: Harlequin ladybirds

Key to identifying whether or not a ladybird is a Harlequin

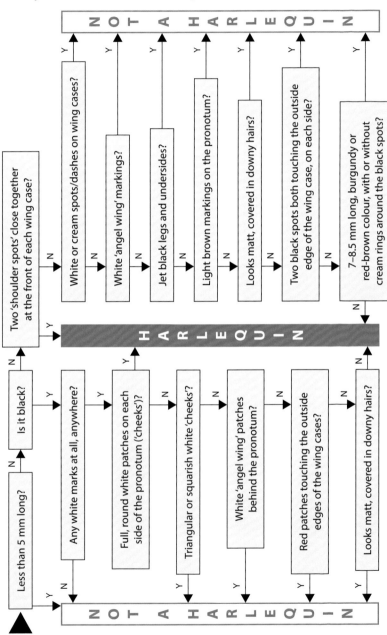

NOT A HARLEQUIN

Two 'shoulder spots' close together at the front of each wing case?

N → White or cream spots/dashes on wing cases? — Y → NOT A HARLEQUIN

N → White 'angel wing' markings? — Y → NOT A HARLEQUIN

N → Jet black legs and undersides? — Y → NOT A HARLEQUIN

N → Light brown markings on the pronotum? — Y → NOT A HARLEQUIN

N → Looks matt, covered in downy hairs? — Y → NOT A HARLEQUIN

N → Two black spots both touching the outside edge of the wing case, on each side? — Y → NOT A HARLEQUIN

N → 7–8.5 mm long, burgundy or red-brown colour, with or without cream rings around the black spots? — Y → NOT A HARLEQUIN

Y (from 'shoulder spots') → **HARLEQUIN**

N (from 7–8.5 mm) → **HARLEQUIN**

Less than 5 mm long? — N → Is it black? — Y → Any white marks at all, anywhere? — Y → Full, round white patches on each side of the pronotum ('cheeks')? — Y → **HARLEQUIN**

Less than 5 mm long? — Y → **NOT A HARLEQUIN**

Is it black? — N → **HARLEQUIN**

Any white marks at all, anywhere? — N → **NOT A HARLEQUIN**

Full, round white patches on each side of the pronotum ('cheeks')? — N → Triangular or squarish white 'cheeks'? — N → White 'angel wing' patches behind the pronotum? — N → Red patches touching the outside edges of the wing cases? — N → Looks matt, covered in downy hairs? — N → **HARLEQUIN**

Triangular or squarish white 'cheeks'? — Y → **NOT A HARLEQUIN**

White 'angel wing' patches behind the pronotum? — Y → **NOT A HARLEQUIN**

Red patches touching the outside edges of the wing cases? — Y → **NOT A HARLEQUIN**

Looks matt, covered in downy hairs? — Y → **NOT A HARLEQUIN**

The identification key (*opposite*) will help you determine whether or not a ladybird is a Harlequin. Start at the ▶ symbol and work your way through the Yes/No (Y/N) questions until you arrive at an identification.

Note that the size of the ladybird is of paramount importance for identification, and the key therefore starts with a question about size.

Whenever you are referring to photographs of ladybirds, beware of white patches caused by reflections/flash/glare.

Remember – unlike Harlequins, some species of ladybird have white patches each side of the central black spot on the front of their wing cases, just behind the pronotum (*right*). Try thinking of this as looking like an angel with white 'angel wings' – that is, a round black body with a little black head in between the white wings. Harlequins never have these white 'angel wing' markings.

Only Harlequins have two black 'shoulder spots' right at the front of the wing cases, always close together, just behind the pronotum (*right*).

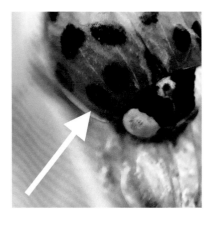

Conspicuous ladybirds of Britain and Ireland

Adonis' ladybird
(*Hippodamia variegata*)
(4–5 mm)

Red with 3–15 (usually 9) black spots. White 'angel wing' markings. White pronotum with distinctive black pattern with two little white dots. Elongated body. Black legs. Found on plants and shrubs in areas of dry soil and sand.

Bryony ladybird
(*Henosepilachna argus*)
(5–7 mm)

Fairly large orange ladybird with a matt appearance due to being covered in short downy hairs, 11 black spots. The orange pronotum has no markings on it. Brown legs. Generally found on the plant white bryony.

Cream-spot ladybird (*Calvia quattuordecim-guttata*) (4–5 mm)

Chestnut-brown with 14 creamy white spots. Maroon-brown pronotum. Brown legs. Found in hedgerows and deciduous trees.

Cream-streaked ladybird (*Harmonia quadripunctata*) (5–6 mm)

Red/orange or salmon-coloured with cream-streaked markings, usually with either 4 or 16 black spots. (Note: There are four bands of spots across the wing cases of the Cream-streaked ladybird, but only three bands on the similarly marked 10-spot ladybird.) White pronotum with 5–9 black spots. Brown legs. Found in coniferous woodland.

Eyed ladybird (*Anatis ocellata*) (7–8.5 mm)

The largest ladybird in Britain. Burgundy or russet, usually with pale rings around its 15 or so black spots. White pronotum with chunky black 'M' mark. Black legs. Found on or near coniferous trees.

Harlequin ladybird (*Harmonia axyridis*) (5–8 mm)

The most common form in Britain is red, orange or yellow with between 0 and 19 black spots. The black forms have either 2 or 4 large red or orange spots on the wing cases. Black and white or cream pronotum. Brown or brownish legs. Found in a wide variety of places, especially urban environments.

Heather ladybird
(*Chilocorus bipustulatus*)
(3–4 mm)

Black all over with a line of 3
tiny red spots on each wing case,
going from one side to the other.
Black legs. Found on heathland
near conifers.

Hieroglyphic ladybird
(*Coccinella hieroglyphica*)
(4–5 mm)

Reddish-brown or pale brown.
Usually has 0–7 black spots/
dashes but can also have slight
grid-like black markings. Black
forms are common. White
'angel wing' markings visible on
all forms. Black legs. Found on
heathland.

Kidney-spot ladybird
(*Chilocorus renipustulatus*)
(4–5 mm)

Black all over with 2 red spots.
The spots are often slightly
kidney-shaped. Black legs. Found
in deciduous woodland.

Larch ladybird (*Aphidecta obliterata*) (4–5 mm)

Light brown or tan-coloured (sometimes with a few brown 'freckles' or a few dark bands down the wing cases). 0–10 dark brown spots/dashes. Pale brown pronotum with pale brown markings. Brown legs. Found in coniferous woodland.

Orange ladybird (*Halyzia sedecimguttata*) (4.5–6 mm)

Bright orange ladybird with 12–16 white spots, yellow markings on its pronotum, jet black eyes and a noticeably translucent flared rim to both pronotum and wing cases. Orange legs. Found in deciduous woodland.

Pine ladybird (*Exochomus quadripustulatus*) (3–4.25 mm)

Small black ladybird with 4 red spots and a noticeable rim to its elytra. The front spots look like a single speech mark or a comma. Black legs. Found in or near a variety of trees or shrubs.

Scarce 7-spot ladybird (*Coccinella magnifica*) (6–8 mm)

Red with 5–11 black spots (usually 7) and 'angel wing' markings. Black pronotum with squarish white 'cheeks'. Very similar to 7-spot ladybird but has a slightly more 'bulging' appearance. Often has larger central spots. Black legs. Found near wood ant nests, usually in heathland and woodland.

To determine whether or not it is a Scarce 7-spot, turn it onto its back and look carefully when it kicks its legs. The 7-spot has a small white triangular mark on each side under the middle pair of legs. The Scarce 7-spot has these marks but also has them under the hindlegs.

Striped ladybird (*Myzia oblongoguttata*) (6–8 mm)

Second-largest ladybird in Britain. Glossy golden brown with cream stripes and spots. Brown and white pronotum. Brown legs. Quite a specialist ladybird found on Scots pine.

Water ladybird (*Anisosticta novemdecimpunctata*) (4 mm)

Reddish in the summer and buff-coloured in autumn and winter, these elongated and flattened little ladybirds have 15–21 black spots. Buff pronotum with 6 black spots. Brown legs. Found in reedbeds and wet marshy areas.

2-spot ladybird (*Adalia bipunctata*) (4–5 mm)

A very variable little ladybird. Usually either red with 2 black spots (*top*) or black with 2 or 4 red spots/patches (*bottom*). As with black forms of the 10-spot, the two front red patches extend to the edges of the wing cases.

Often the black pronotum has tiny white crescents on its outer edges. Black legs. Found in a wide variety of places including urban areas.

5-spot ladybird (*Coccinella quinquepunctata*) (4–5 mm)

Red with 5–9 black spots (usually 5) and white 'angel wing' markings. Black pronotum with squarish white 'cheeks'. Black legs. Only found on unstable river shingle.

7-spot ladybird (*Coccinella septempunctata*) (5–8 mm)

This very common red ladybird has 0–9 black spots (usually 7) and white 'angel wing' markings. Black pronotum with squarish white 'cheeks'. Black legs. Found in a wide variety of places, but often on low-growing plants.

10-spot ladybird (*Adalia decempunctata*) (3.5–4.5 mm)

This little ladybird comes in three exceedingly different forms. The first (*right*) is black/dark brown/purple with two orange/yellow/red flashes that extend

to the elytra edges. The second form (*above*) is red/orange/yellow with 0–15 dark brown/black spots. The third form (*right*) is brown/black chequered with a cream/yellow/red grid – this form used to be called the 'bed-sock ladybird'. Brown legs (a very helpful identification tool). Found on hedgerows and deciduous trees.

11-spot ladybird (*Coccinella undecimpunctata*) (4–5 mm)

This slightly elongated red ladybird has white 'angel wing' markings and 7–11 (usually 11) black spots. Black pronotum with small, squarish white 'cheeks'. Black legs. Found mainly in sandy and coastal habitats.

13-spot ladybird (*Hippodamia tredecimpunctata*) (5–7 mm)

An uncommon red ladybird with an elongated and flattened body with 7–15 (usually 13) black spots. The white pronotum has a distinctive solid black central marking with two small side dots. Legs are golden brown with black 'thighs'. Very rare, found mainly in the south of England.

14-spot ladybird (*Propylea quattuordecimpunctata*) (3.5–4.5 mm)

Yellow with usually 14 distinctive rectangular black spots that often fuse together to resemble a clown's face or an anchor. Distinctive pronotum markings that can look like the top of a hippo's head in the water! Brown legs. Found in a diverse range of habitats.

16-spot ladybird (*Tytthaspis sedecimpunctata*) (3 mm)

A small beige ladybird with a distinctive black line down its back and (usually) 16 black spots that often fuse together near the outside edges, forming a zigzag line. Brown legs. Found in grassland, near the ground.

18-spot ladybird (*Myrrha octodecimguttata*) (4–5 mm)

A chestnut-brown ladybird with 14–18 (usually 18) cream/white spots. Spots fuse together at the top of the wing cases where they meet the pronotum, creating a curved pattern. Cream/white pronotum with brown markings. Brown legs. Found in conifer woods.

22-spot ladybird (*Psyllobora vigintiduopunctata*) (3–4 mm)

A small yellow ladybird with 20–22 black spots. The yellow or cream pronotum has 4 black spots with a little black triangle in the middle. Brown legs. Found on low vegetation.

24-spot ladybird (*Subcoccinella vigintiquattuorpunctata*) (3–4 mm)

A small orange-red ladybird covered in soft, fine downy hairs that give it a matt appearance. It also has black spots or marks on its pronotum. Its 0–24 (usually 20) black spots often merge together in places. Orange-red legs. Found on grasses and other low-growing plants.

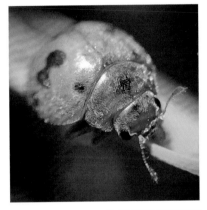

Top row, left to right: Larch, 22-spot and Bryony ladybirds

Main photo: Mating 10-spot ladybirds

Middle right: Eyed ladybird, 14-spot ladybird

Bottom row: 7-spot and two Harlequin ladybirds

Other Harlequins

In addition to the three forms of Harlequin ladybird currently found in Britain (*Harmonia axyridis* f. *succinea*, f. *spectabilis*, and f. *conspicua*), there are four additional forms found in Europe. Some of these forms have occasionally turned up in Britain and Ireland.

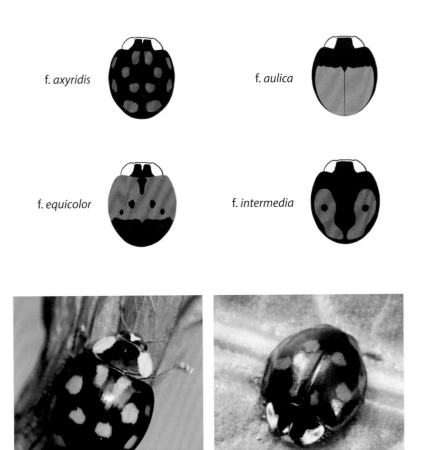

f. *axyridis*

f. *aulica*

f. *equicolor*

f. *intermedia*

Above: f. *axyridis* is black with red/orange spots

Above and left: The rear third of f. *equicolor* is always black, with the front two-thirds being red/orange with or without black spots

Above: The rear three-quarters of f. *aulica* is red-orange or honey-coloured, with the front quarter black

Above: f. *intermedia* is black with two elongated red/orange
patches, with or without black spots within the patch

How to record your sightings

The UK Ladybird Survey is very interested in getting records of as many ladybird sightings as possible and encourages everyone to record any ladybirds they see. This can be done via the UK Ladybird Survey online (www.coleoptera.org.uk/coccinellidae), or via iRecord, the iRecord app or the European Ladybirds app. Submitting photographs as well is particularly useful in order to help with verification.

Glossary

Abdomen – Third section of an insect's body (contains the digestive tract and reproductive organs).

Aggregation – A clustering together of several ladybirds.

Antennae – A pair of sensory appendages on the heads of insects, sensitive to vibrations, touch and in some species sound and smell (singular: antenna).

Aphids – Greenfly or blackfly.

Conspicuous ladybirds – Black or brightly coloured ladybirds, 3–8.5 mm in length.

Elytra – The hard, outer wing cases, also known as forewings (singular: elytron).

Form – The name given to a colour form or morph of a particular species (abbreviation: f.).

Head – First section of an insect's body (where the antennae, compound eyes and mouthparts are found).

Host – A living organism on or in which a parasite lives.

Inconspicuous ladybirds – Small, indistinct and slightly hairy species that may not easily be recognised as ladybirds, 1.2–4 mm in length.

Instar – The stage between two moults of an insect larva.

Intraguild predation – Predation by one species on another with which it shares resources such as food. The killing and sometimes eating of a potential competitor of a different species.

Larva – The second immature stage of insects that undergo complete metamorphosis, i.e. between egg and pupa (plural: larvae).

Melanic – Black colour form of an adult ladybird.

Metamorphosis – The changes that take place during an insect's life as it goes from egg to adult.

Mildew – A type of white fungus often seen growing on plant leaves.

Moult – The shedding of the skin or outer covering of the body.

Parasite – An organism that lives in or on another, obtaining resources from, but not killing the host.

Parasitoid – An organism that lives as a parasite, but usually kills the host.

Pronotum – Hard protective plate which protects the head and thorax (plural: pronota).

Pupa – Final immature stage of insects that undergo complete metamorphosis, i.e. between larva and adult (plural: pupae).

Reflex bleeding – Defensive reaction in which reflex blood is exuded from leg joints or pores.

Reflex blood – Noxious yellow-coloured secretion produced by ladybirds (and some other insects) when threatened.

Scoli – Distinctive fleshy branched spines on the body of a ladybird larva (singular: scolus).

Thorax – Middle section of an insect's body, behind the head and bearing the legs and wings.

Further reading and references

Ando, T. and Niimi, T. 2019. Development and evolution of color patterns in ladybird beetles: A case study in Harmonia axyridis. *Development, Growth and Differentiation* 61: 73–84.

Brown, P.M.J. and Roy, H.E., 2018. Native ladybird decline caused by the invasive Harlequin ladybird *Harmonia axyridis*: evidence from a long-term field study. *Insect Conservation and Diversity* 11: 230–239.

Brown, P.M.J., Adriaens, T., Bathon, H. *et al.* 2008. *Harmonia axyridis* in Europe: spread and distribution of a non-native coccinellid. *BioControl* 53: 5–21.

Brown, P.M.J., Roy, D.B., Harrower, C., Dean, H.J., Rorke, S.L. & Roy, H.E. 2018. Spread of a model invasive alien species, the Harlequin ladybird *Harmonia axyridis* in Britain and Ireland. *Scientific Data* 5: 180239.

CABI. 2019. Invasive Species Compendium: *Harmonia axyridis* (Harlequin ladybird). www.cabi.org/isc/datasheet/26515.

Ceryngier, P., Nedvěd, O., Grez, A.A. *et al.* 2018. Predators and parasitoids of the Harlequin ladybird, *Harmonia axyridis*, in its native range and invaded areas. *Biological Invasions* 20: 1009–1031.

Haelewaters, D., Minnaar, I.A. and Clusella-Trullas, S. 2016. First finding of the parasitic fungus *Hesperomyces virescens*

(Laboulbeniales) on native and invasive ladybirds (Coleoptera, Coccinellidae) in South Africa. *Parasite* 23: 5.

Holloway, G.J., de Jong, P.W., Brakefield, P.M. and de Vos, H. 1991. Chemical defence in ladybird beetles (Coccinellidae). I. Distribution of coccinelline and individual variation in defence in 7-Spot ladybirds (*Coccinella septempunctata*). *Chemoecology* 2: 7–14.

Lombaert, E., Guillemaud, T., Cornuet, J.M. *et al.* 2010. Bridgehead effect in the worldwide invasion of the biocontrol Harlequin ladybird. *PLoS One* 5 (3): e9743.

Majerus, M.E.N. 1994. *Ladybirds*. New Naturalist 81. Harper Collins, London.

Majerus, M.E.N., Roy, E.H., Brown, P.M.J., Poland, R.L. and Shields, C. 2006. *Guide to Ladybirds of the British Isles*. Field Studies Council. A laminated pull-out chart.

Majerus, M.E.N., Ware, R. and Majerus, C. 2008. *A Year in the Lives of British Ladybirds*. AES Bug Club, Orpington.

Majerus, M.E.N., Roy, E.H. and Brown, P.M.J. 2016. *A Natural History of Ladybird Beetles*. Cambridge University Press, Cambridge.

Martin, N.A. 2016 (revised 2018). Harlequin ladybird – *Harmonia axyridis*. New Zealand Arthropod Collection Factsheet. nzacfactsheets.landcareresearch.co.nz.

Mezőfi, L. and Korányi, D. 2017. The colour pattern forms of the Harlequin ladybird (*Harmonia axyridis*, Pallas 1773) in Hungary and the ecological aspects of its polymorphism – A harlekinkatica (*Harmonia axyridis* Pallas 1773) színváltozatai Magyarországon éspolimorfizmusának ökológiai vonatkozásai. *Növényvédelem* 53: 193–205.

Roy, H.E. and Brown, P.M.J., 2015. Ten years of invasion: *Harmonia axyridis* (Pallas) (Coleoptera: Coccinellidae) in Britain. *Ecological Entomology* 40: 336–348.

Roy, H.E. and Brown, P.M.J. 2018. *Field Guide to the Ladybirds of Great Britain and Ireland*. Bloomsbury, London.

Roy, H.E., Brown, P.M.J., Comont, R.F., Poland, R.L. and Sloggett, J.J. 2013. *Ladybirds*, 2nd edition. Naturalists' Handbook 10. Pelagic Publishing, Exeter.

Roy, H.E., Brown, P.M.J., Adriaens, T. *et al.* 2016. The Harlequin ladybird, *Harmonia axyridis*: global perspectives on invasion history and ecology. *Biological Invasions* 18: 997–1044.

Saito, K., Nomura, S., Yamamoto, S., Niiyama, R. and Okabe, Y. 2017. Investigation of hindwing folding in ladybird beetles by artificial elytron transplantations and microcomputed tomography. *Proceedings of the National Academy of Sciences of the USA* 114: 5624–5628.

Southampton Natural History Society. 2005. *Ladybirds of Southampton*. sotonnhs.net/wp-content/uploads/Documents/Survey-Ladybirds.pdf.

Wiorek, M. 2018. Inwazyjny arlekin: biedronka azjatycka. *Wszechświat, Pismo Przyrodnicze* 119 (10–12): 234–243.

Photographic credits

With the exception of those listed below, the photographs have all been taken by the author and are copyright of Helen Boyce. Thanks to the following for providing additional photographs:

p. 8 Aggregation of Harlequins © Gilles San Martin

p. 19 2-spot larva © Yvonne Couch

p. 19 11-spot and Adonis' ladybird larvae © Gilles San Martin

p. 20 13-spot, 16-spot, 18-spot, Hieroglyphic and 5-spot ladybird larvae © Gilles San Martin

p. 21 Water ladybird larva © Yvonne Couch

p. 21 Heather ladybird larva © Gilles San Martin

p. 32 Newly emerged 22-spot ladybird © Yvonne Couch

p. 40 Spider and Harlequin © Ann Miles

p. 41 Parasitised 7-spot ladybird © Stephen Loftin

p. 83 Heather ladybird © Charles J Sharp

p. 86 Mating 2-spot ladybirds © Gail Hampshire

p. 87 5-spot ladybird © Gilles San Martin

p. 88 11-spot ladybird © Sandy Rae

p. 89 13-spot ladybird © Gilles San Martin

p. 93 *Harmonia axyridis* f. *axyridis* on purple flower © Thijs de Graaf

p. 94 *Harmonia axyridis* f. *equicolor* (×2 top row) © Stanislav Krejčík

p. 94 *Harmonia axyridis* f. *equicolor* (second row) © Sharon Wegh

p. 94 *Harmonia axyridis* f. *aulica* (×2) © Nikola Rahmé

p. 95 *Harmonia axyridis* f. *intermedia* on leaf (bottom right) © Bernard Fransen

Index of non-Harlequin photos

Notes